I0409145

I would like to express my special thanks to one of my very good friends Mrs. Fadhilah Abdhullah (a passout of Lincoln University), Dr. Arun Kumar Sinha (Maternal Grandfather), Dr. Amita Sinha, Mother and Dr. Arun Kumar Mehta, Father for their continuous love, support and motivation to write this book on a step towards better future of our Earth with Ecotourism and Eco-friendly clothes.

PREFACE

This book has been written to provide knowledge towards protecting the environment and conservation of natural resources with the help of sustainable traveling and promoting slow fashion. I also would like to provide respect to my pet dog Toughie who lived with us for 18 years which is a long and memorable period. I got this motivation to write this book when I came from New Zealand (Christchurch) where I saw beautiful landscapes and birds. This is why I am writing this book as I am a nature lover and we should always respect what it has given to us.

.

CHAPTER-1

Ecotourism was first coined in 1983 by "Hector Ceballos Lascurain", a Mexican architect turned into an Environmentalist & defined ecotourism in the year 1995 as "tourism that consists of travelling to relatively undisturbed or uncontaminated natural areas with the specific object of studying, admiring and enjoying the scenery and its wild plants and animals, as well as any existing cultural manifestations (both past and present) found in these areas".

In other words, ecotourism is called environment-friendly tourism that deals with various types of issues and factors related to human impact on natural areas and their resources caused due to tourism and seeks to address them how Ecotourism has created positive impacts, whereas Tourism has created negative impacts on nature. In this book we have disscussed both positive impact of Ecotourism and negative impacts of Tourism.

The positive impacts of ecotourism are: It reduces hunger and poverty. Creates awareness towards the protection of tourist spots that are under threat by the increase of human activities. It educates people to conserve natural resources and wildlife species. It promotes the idea of sustainable business and marketing.

Ecotourism contributes to economic development and generates employment opportunities. Ecotourism helps in a better understanding of the culture and traditions of other communities. It provides financial benefits for conservation and led to the empowerment of women communities. Ecotourism is the best way of preventing the illegal trafficking of animal body parts. Minimizes human impact on marine and coastal environments.

* * *

Case study on Positive Impacts of Ecotourism

Case studies we have considered to mention the positive impacts of Ecotourism are below:
1) Ecotourism case study on how it affects awareness and attitudes but not conservation behaviors: A case study on Grand Riviere, Trinidad.

2) Exploring the Impact of Ecotourism on Indonesia's Environment.

* * *

1) Ecotourism case study on how it affects awareness and attitudes but not conservation behaviors: A case study on Grand Riviere, Trinidad

Aim of the case study: This case study aims at looking at the impact of Ecotourism on awareness attitudes and conservation behaviors of the local communities in the Grand Riviere, Trinidad. The methods used in this case study were Rapid Rural Appraisal and questionnaire-based interviews to investigate the issue in this village of Grand Riviere, Trinidad.

*I*ntroduction to the case study: Ecotourism is one of the powerful tools for creating awareness towards conservation of natural resources and promoting responsible way of travelling to tourist spots. In this case study, according to International Tourism Society Ecotourism is defined as the responsible way of travelling to natural areas that conserves the environment and sustains the well-being of local people

(Wood,2002). Tourism can bring significant financial benefits to areas supporting charismatic wildlife (Walpole & Leader-Williams, 2002; Adams & Infield, 2003; Lorimer, 2009) and thus can be a relatively cheap method of facilitating both development and conservation (Cater & Goodall, 1997; UNWTO, 2006). Ecotourism grew three times faster than the overall tourism industry in 2004, according to The International Ecotourism Society. This growth reflects the increasing importance of sustainable and environmentally conscious travel. Ecotourism is not a panacea for all conservation problems (Weaver, 1998; Kruger, 2005). Ecotourism can sometimes benefit both people and threatened species, many ecotourism projects are likely to fail in achieving their goals for either conservation (Yu et al., 1997) or development (Bookbinder et al., 1998). Therefore, we need to gain a better understanding of the impacts of ecotourism because there are no simple alternatives.

Community-based conservation has also seen as a limited success (Adams et al., 2004). One way in which tourism could demonstrate its value is by improving local awareness and attitudes toward conservation. Community-based conservation initiatives assert that providing financial or livelihood benefits, along with increased participation, will undoubtedly encourage positive attitudes towards conservation (Spiteri & Nepal, 2006). For instance, local community attitudes towards Serengeti National Park were negatively affected by perceived costs, but participation in a community-based project significantly improved their perception (Kideghesho et al., 2007). It is firmly believed that education programs can influence attitudes, despite potential variations (Brossard et al., 2005). Creating positive attitudes toward conservation is important when other methods of behavior change are ineffective. Increasing awareness is necessary and can predict conservation behaviors (Maibach, 1993; Beedell & Rehman, 2000). Psychologists agree that awareness and attitudes can be important predictors of behavior in combination with other factors (Ajzen, 2005). Attitudes are influenced by pre-existing values (Schultz, 2001) and by issues and processes that are not obvious to outsiders (Allendorf et al., 2006), and thus specifying the attitudes most appropriate to conservation outcomes is not clear (Saunders et al., 2006). Ecotourism initiatives (Kiss, 2004) have unequivocally stated that positive attitudes alone are inadequate to induce behavioral change. The detailed and poorly understood correlation between attitude change and behaviors impacting conservation outcomes necessitates thorough verification in attitude studies to ensure anticipated behavioral changes. This study aims to explore the impact of ecotourism on people's awareness and attitudes towards nature and its conservation. The research will compare perceptions of the Trinidad piping-guan (Pipile Pipile) and the Leatherback Turtle, which have been the focal points of various conservation initiatives. Additionally, it resolves analyze the effects of direct ecotourism benefits on attitudes and awareness.

Study area and Methods: The Caribbean has an ecology that is more similar to the adjacent South American mainland than the rest of the Caribbean. The only bird that is unique to this region is the Trinidad piping-guan, known locally as the Pawi. It is listed as Critically Endangered on the IUCN Red List (IUCN, 2008). Hunting is believed to be a significant factor in its decline (Brooks, 1999), but the socio-economic context of this decline is poorly understood (James & Hislop, 1998). The village of Grande Riviere, located on the impoverished and remote northeast coast of Trinidad (10.5N, 61.5W), is a crucial site for spotting the Pawi, as it is one of the few reliable places to do so (Hayes, 2002). Grande Riviere has the most tourist development of any village in this area. There are two small hotels next to the beach, one opened in 1993 and the other in 2000, as well as a third constructed in 2005. Additionally, there are a few guest houses in the village. By 2006, about 10,000 tourists visited the village annually, with the majority coming to see the turtles (Harrison, 2007). In 1989, there was a high number of Leatherback Turtles being killed on the northeast coast. It was common for people to eat Turtle meat and eggs due to cultural traditions (James & Fournier, 1993). However, since the start of ecotourism activities in 1992-1993, the consumption of Turtle meat has become less common (Harrison, 2007). During the turtle laying and hatching season, access to the beach is prohibited between 18:00 and 06:00 without a permit and accompanying authorized guide. The success of this initiative is due to the commitment of local people (Onwuka, 2004). Similar schemes have failed elsewhere in Trinidad (James & Fournillier, 1993). Conservation legislation includes an annual 6-month hunting ban for all game species. Some species, such as the Pawi, are protected year-round. In October 2004, Matura National Park existed established next to the village and designated as an Environmentally Sensitive Area. The Forestry Division has organized various educational activities, often in schools, and has launched a long-term education campaign for the Pawi in 2005. The impact of these initiatives on local attitudes and behaviors has not been tracked.

Methods: carried out to study the behavior of the Grande Riviere village are rapid rural appraisal (RRA) and questionnaire-based interviews. The techniques vary, but they underline visual approaches that encourage community participation (Kapila & Lyon, 1994). In this study, they were used to understand the general context of village life, as well as perceptions, use, and attitudes towards local natural resources. A men's group, a women's group, a mixed group and a children's group (each of 3–6 members) separately discussed similar issues, with some repetition of exercises between groups to validate conclusions. A timeline of the village's history, types of land and their uses, natural resource utilization, seasonality of natural resources, food sources, livelihood ranking, and wildlife conservation efforts. Questionnaire was developed based on the information gathered from the RRA. It was created to gather data on knowledge, usage, and attitudes toward local natural resources, as well as household socio-economic characteristics. Trinidad is an English-speaking country, but a Trinidadian research assistant was utilized to avoid any issues related to accent or cultural differences. Open-minded, diverse responses were used to avoid bias and later categorized for analysis.

Results: **Awareness and attitudes towards Turtles** and Pawi, most of the respondents who were interviewed conservation of Pawi (Trinidad piping guan) was not mentioned and not given more attention, as for the conservation of Leatherback Turtle. The awareness of conservation initiatives for turtles was higher than for Pawi. More people remembered education efforts for turtles, and in a knowledge test, more respondents could recognize Turtles than Pawi. Additionally, those naming Pawi were more likely to have a higher knowledge score (excluding points for Pawi) suggesting that knowledge of Pawi may be a result of education programs rather than direct observation. **Effect of tourism benefits on attitudes and awareness**: There were two indicators of direct benefits from Ecotourism: Households with members working as tour guides or in hotels had significantly better knowledge and were more likely to identify turtles as a threatened species. Tourism had no impact on local support for conservation or ecotourism. 51 out of 52 respondents supported ecotourism. Respondents from tour-guide or hotel-worker households showed no bias. Retirees were excluded to avoid bias. Perceptions and Behaviours: Hunting and wild meat consumption reflect conservation awareness. The community perceives hunting as a significant threat to wild animals, with a bias towards animals hunted for meat. In informal discussions, hunting is often cited as the primary reason for the vulnerability of these species. There was a strong preference for wild meat over domestic meat (50 out of 52 preferred wild meat), although not everyone could access or afford it. Additionally, most respondents did not eat it often, typically consuming it less than once per month. Popularity often influences price. For example, meat from the manicou Didelphis marsupialis can sell for up to USD 18 per kilogram, while chicken meat costs about USD 1 per kilogram. Research showed that Pawi meat was not highly valued by hunters but would likely be captured if encountered during hunting.

Conclusion: Ecotourism has the potential to positively influence awareness and attitudes toward nature, making it an important tool for conservationists. However, to ensure that attitudes translate into behaviour, there must be a focus on personal behavioural change. Studies should take into account the potential mismatch between awareness and attitudes towards conservation issues and behaviours.

❋ ❋ ❋

A brief introduction on Trinidad Piping Guan

Scientific Name of Trinidad Piping Guan: Aburria pipile

Population: 50-250 individuals left in the wild.

IUCN status: Critically Endangered

CITES: Appendix1 (includes species threatened with extinction)

Weight: 2,500–3,300 grams

Height: 24-28 inches (60.96 cm to 71.12 cm)

Types of Piping Guan:

- Blue throated Piping Guan
- Trinidad Piping Guan
- Red throated Piping Guan
- Wattled Guan
- Black Fronted Piping Guan

There are five different types of Piping Guan that are found in wildlife.

Man, who coined the genus name is German naturalist Gregor Johann in the year 1830.

Food and Habitat: The Trinidad piping guan, also known as the Pawi, is a bird species native to the island of Trinidad. It primarily eats fruits such as palm fruits, wild figs, mangos, guavas, cherries, and berries, as well as the fruits of forest trees and shrubs. During times when its preferred fruits are scarce, it may also consume seeds, leaves, buds, and flowers. Insects and other invertebrates are occasionally eaten, but they are not a significant part of its diet.

The Trinidad piping guan lives in lowland and foothill forests in Trinidad, typically below 900 meters (3,000 feet) above sea level. It prefers mature, undisturbed primary forests with a closed canopy and a well-developed understory. These birds are often found in areas with large, tall trees that provide abundant fruit resources, such as evergreen and semi-evergreen forests. The presence of water sources, such as rivers or streams, nearby also seems to be important for their habitat selection.

The piping guan typically forages for food on the forest floor, in trees, or in shrubs, depending on the availability of its preferred fruits. It requires large, contiguous areas of forest to find sufficient food resources and suitable nesting sites.

The Trinidad piping guan plays an important ecological role in its habitat as a seed disperser, contributing to the regeneration and maintenance of the forest ecosystem. However, its dependence on specific forest habitats makes it vulnerable to habitat loss and fragmentation. Conservation efforts focus on protecting and restoring their natural forest habitats, as well as reducing threats such as hunting and trapping, which have contributed to their vulnerable status according to the IUCN Red List.

Threats to Trinidad Piping Guan: The Piping Guan, scientifically known as Aburria pipile, is a bird species endemic to Trinidad and Tobago. This large, turkey-like bird is facing a myriad of threats that risk its existence in the wild. Habitat loss, primarily due to deforestation for timber extraction and agricultural expansion, poses a significant challenge to the Piping Guan's survival. The destruction of its natural habitats disrupts its breeding and foraging grounds, leading to a decline in population.

Additionally, illegal activities such as hunting and trapping for the pet trade have further exacerbated the species' vulnerability. Unregulated hunting and capture for the wildlife trade have caused a decline in Piping Guan numbers, threatening the genetic diversity and overall health of the population. Moreover, introducing non-native species and diseases poses a grave risk to the birds, as they lack natural defences against these threats.

Conservation efforts aimed at safeguarding the Piping Guan are multifaceted. Initiatives for habitat protection focus on preserving and restoring the bird's natural ecosystems. This includes the establishment of protected areas and the promotion of sustainable land-use practices to reduce deforestation. Moreover, stringent law enforcement and regulations are essential to combat illegal hunting and trading of the species.

Raising awareness about the plight of the Piping Guan and the importance of its conservation is another pivotal aspect of the efforts. Through education and outreach programs, local communities and the public are informed about the significance of preserving the species and its habitat. Engaging stakeholders in conservation activities promotes a shared responsibility for protecting the Piping Guan and its ecosystem.

In conclusion, conserving the Piping Guan in Trinidad and Tobago requires comprehensive and concerted efforts. By addressing the threats of habitat loss, illegal activities, and susceptibility to introduced species and diseases, there is hope for the continued survival of this iconic bird species.

Why are these birds (Piping Guan) gets illegally

traded and what are the ways of preventing the same:
The Piping Guan, scientifically known as is illegally
traded in the pet trade for several reasons:

The Piping Guan, scientifically known as is illegally traded in the pet trade for several reasons:

1. Unique Appearance: Piping Guans are magnificent beautiful birds with vibrant blue plumage, a red bill, and a distinctive white crest. Their colourful appearance makes them highly desirable to bird enthusiasts and collectors who value rare and exotic pets.

2. Distinctive Vocalizations: These birds have a unique and loud call, described as a high-pitched, piping whistle, which gives them their name. Some people find their vocalizations intriguing and may seek to own a Piping Guan to experience this distinctive sound.

The vulnerability and rarity of the Piping Guan can increase its appeal to collectors seeking status symbols or exclusivity.

4. Lack of Legal Alternatives: The Piping Guan is a vulnerable species with a limited range and small population size. They are not commonly bred in captivity, so the only way to obtain one is through illegal trapping and trading, which fuels the demand in the black market.

5. Cultural Beliefs: In some regions, there may be cultural beliefs or traditions associated with owning or keeping certain bird species as pets. This could drive the demand for Piping Guans, even if it is not legal or sustainable.

6. Financial Gain: The illegal trade of wildlife, including the Piping Guan, is often driven by financial motives. Trappers and traders can make significant profits by selling these rare birds on the black market, despite the ethical and ecological developments.

❋ ❋ ❋

It's important to note that the trade of Piping Guans is dangerous to their already vulnerable populations and can have severe ecological impacts. It disrupts their natural behaviour, causes stress, and removes individuals from their vital role in maintaining the health and diversity of their ecosystems.

Conservation efforts focus on educating local communities about the negative effects of the pet trade, enforcing laws and regulations, and promoting sustainable alternatives to generate income, such as birdwatching tourism, that can benefit both people and the conservation of the Piping Guan.

✱ ✱ ✱

Step to be taken for conserving Trinidad Piping Guan:

Trinidad piping guan (Pipile pipile), also known as the paw or the white-headed piping guan. This bird species is endemic to the island of Trinidad and is considered vulnerable by the IUCN Red List due to its small and declining population. Here are some ways to conserve the Trinidad piping guan:

1. Habitat Preservation and Protection:
   - Protect and preserve the bird's natural habitat, mainly the lowland and foothill forests of Trinidad. This includes preventing deforestation, maintaining forest connectivity, and restoring degraded habitats.
   - Establish and effectively manage protected areas specifically for the conservation of the Trinidad piping guan and its habitat.

2. Research and Monitoring:
   - Conduct comprehensive surveys and research to better understand the bird's ecology, behaviour, and population dynamics. This information is crucial for developing effective conservation strategies.
   - Implement long-term monitoring programs to track the

population status, distribution, and habitat use of the Trinidad piping guan.

3. Address Threats: - Control and mitigate threats such as hunting and trapping, which significantly contribute to the decline of the species. Increase law enforcement and raise awareness about the legal protections in place for the bird.

- Reduce the risk of collisions with man-made structures such as power lines and vehicles. Implement measures such as bird-safe power line designs and speed limits in areas where the birds are known to cross roads.

4. Community Involvement and Education:

- Engage local communities in conservation efforts by raising awareness about the importance of protecting the Trinidad piping guan and its habitat. Educate the public about the bird's ecological role and the threats it faces.

- Encourage ecotourism that focuses on responsible birdwatching and promotes the appreciation of the Trinidad piping guan, generating economic benefits for local communities.

5. International Cooperation:

- Collaborate with international organizations and institutions working on bird conservation, particularly those with expertise in Neotropical bird species. Exchange knowledge, share best practices, and seek support for conservation initiatives.

6. Captive Breeding and Reintroduction:

- Establish captive breeding programs to ensure the survival of the species and to produce individuals for potential reintroduction into suitable habitats.

- Develop and implement reintroduction programs based on thorough research and planning, taking into account habitat availability, potential release sites, and the involvement of local communities.

7. Policy and Legal Framework:

- Advocate for strong national and local policies that protect the

Trinidad piping guan and its habitat. Ensure that relevant laws and regulations are effectively enforced.

- Encourage the inclusion of the bird's conservation in national biodiversity strategies and action plans.

By implementing these conservation measures, it is possible to protect and restore the population of the Trinidad piping guan, ensuring its long-term survival in its natural habitat. Collaboration between government agencies, non-governmental organizations, local communities, and international partners is key to the success of these conservation efforts.

❋ ❋ ❋

2) Environmental effects of Ecotourism in Indonesia

What is Ecotourism and its development in the Indonesia: The definition of ecotourism was introduced by the International Ecotourism Society (IES) in 1990 (Fandeli, 2000), which states that ecotourism is a form of travel into natural areas aimed at conserving the environmental resources and preserving biodiversity and improving the local community's livelihood. Tourist activities can be regarded as ecotourism if it has met three dimensions: (1) conservation dimensions, namely tourism activities are helping local conservation efforts with minimum negative impacts, (2) academic dimensions, namely the tourists who follow the activities of these tours will get details about ecotourism, unique local biological and sociocultural lives, and (3) social dimensions, the local people who have been key actors in implementing any tourism activities (Hafild, 1995). Ecotourism, based on these criteria, is a form of responsible tourism that focuses on visiting undisturbed natural areas or areas managed according to specific rules in order to enjoy and appreciate the ecosystem services and traditional cultures that support conservation. It involves educational elements, has a low impact on the socio-economic aspects, and actively involves local communities. Ecotourism in Indonesia emphasizes nature and cultural conservation while benefiting local communities. Its key principles include sustainability, conservation, community involvement, education, interpretation, and responsible tourism practices. Indonesia's rich biodiversity and cultural heritage are showcased and preserved through ecotourism, providing economic benefits to local communities. The government has designated specific areas as ecotourism destinations and is actively promoting sustainable tourism practices in these areas.

Development of Ecotourism in Indonesia: Ecotourism in Indonesia focuses on conserving nature and culture while also supporting local communities. Its main principles include sustainability, conservation, community participation, education, interpretation, and responsible tourism practices. Indonesia's diverse biodiversity and cultural heritage are highlighted and protected through ecotourism, bringing economic advantages to local communities. The government has identified specific areas as ecotourism destinations and encourages sustainable tourism practices. In Indonesia, tourism became important in 1995 with a seminar and workshop organized by Pakta Indonesia and WALHI in Bogor. Ecotourism development was initially driven by NGOs, focusing on environmental preservation, economic development, and empowering local communities sustainably. Constraints include forest logging, uncontrolled mining, illegal hunting, flood disasters, social conflicts, and environmental insecurities. With rich biodiversity and cultural attractions, Indonesia has the potential to earn $950 million from ecotourism, with the United States as a potential market. Below are five fundamental principles of ecotourism for implementation in Indonesia:

- Support nature conservation programs.
- Involve local communities.
- Provide economic benefits to tourism communities.
- Preserve social values, local culture, and religious communities.
- Comply with tourism and nature conservation regulations (Anonymous, 1997; Anonymous, 1998).

To promote ecotourism in Indonesia, it's important to involve local communities by building partnerships, including them in planning and decision-making, respecting their knowledge, and providing business and job opportunities.

Exploring Indonesia's Enchanting Tourism Destinations: Indonesia, located in Southeast Asia, welcomes visitors with a wide range of diverse and stunning tourism destinations. From natural marvels to vibrant cultures, the country offers a multitude of experiences. Let's delve into some of the potential tourism destinations that capture the essence of this remarkable nation:

Ø Bali:
Ø Known as the Island of Gods, Bali enthrals travellers with its natural beauty and rich cultural heritage. From lush rice terraces in Ubud to the breathtaking Mount Agung at sunrise, Bali presents a harmonious fusion of nature and tradition. The island also invites exploration of ancient temples, traditional dances, and the enchanting sounds of gamelan music. Additionally, Bali is home to yoga studios and wellness centres, perfect for spiritual retreats.

Ø Borobudur Temple, Java:
Ø Visit the magnificent Borobudur Temple, a UNESCO World Heritage Site in Java, dating back to the 9th century. The ancient Buddhist monument showcases intricate stone carvings depicting the life and teachings of Buddha. Witnessing the sunrise or sunset over the lush landscape from the top of Borobudur is a truly magical experience.

Ø Komodo National Park:
Ø Venture to Komodo National Park for a unique adventure where you can come face to face with the fascinating Komodo dragons. The park also boasts natural beauty with crystal-clear waters, vibrant coral reefs, and stunning landscapes, offering wildlife enthusiasts an unforgettable experience.

Ø Raja Ampat, West Papua:

Ø Explore the diverse marine ecosystem of Raja Ampat, renowned for its rich marine life and coral formations. Above the water, the region features picturesque islands with lush greenery and traditional villages, providing a culturally rich experience.

Ø Lake Toba, Sumatra: Visit Lake Toba, a tranquil destination formed by a massive volcanic eruption. This expansive lake, nestled amidst lush hills and traditional Batak villages, offers a peaceful retreat. Visitors can take boat rides on the serene waters and admire the breathtaking views from the surrounding highlands.

Ø Discover vibrant cities such as Jakarta, Bandung, and Surabaya, each offering its unique charm and attractions. The country also promises lively festivals, traditional arts, and delectable cuisine that reflect its rich cultural heritage. Indonesia's diverse ethnic groups and rich cultural heritage offer a truly unique experience at every turn in the country.

Potential Tourism Objects: All tourism objects and attractions consist of three basic elements (Nyoman, 1990): 1. Places: special locations that tourists can visit.

2. Signs or symbols: visible indicators of the high value of tourism.

3. Physical borders: limitations that define the physical objects or environmental attractions. The carrying capacity of ecotourism is not only limited to the number of visits, but also includes other aspects such as:

1. Ecological capacity - the ability of the natural environment to meet the needs of tourists.

2. Physical capacity - the ability of supporting facilities and infrastructure to meet the needs of tourists.

3. Social capacity - the ability to accept tourism destination sites without negative impacts on local communities.

4. Economic capacity - the ability to accommodate destination commercial efforts while still benefiting the local economy.

Carrying Capacity of Tourism: The classification of tourists according to their preferences for enjoying tourist attractions in a specific location and time can provide valuable information on the tourism carrying capacity. In simpler terms, tourism carrying capacity refers to the maximum number of tourists that can visit a specific tourist attraction within a given area and time period (Soemarwoto, 1997, in Lubis, 2006). Tourism carrying capacity refers to the number of tourists visiting an attraction within a specific area and time. It is determined by the destination and influenced by the natural environment and socio-cultural values for long-term preservation (Soemarwoto, 1997; Lubis, 2006). Their leisure activities are expected to aid in the recovery of physical and psychological disorders. Environmental attractions comprise various interconnected biological and physical components, including endemic flora and fauna also its wildlife habitat. The physical components for tourism include topography, soil qualities, climate, and supporting facilities (Douglass, 1978; Lubis, 2006). The distinctive features and unique specifications, along with supporting infrastructures for tourists and natural attractions, are essential for offering high recreational value. This could include beautiful views of mountain landscapes, rivers, beaches, dunes, forests, and other natural beauty. Distinctive features and specifications of uniqueness, availability of supporting infrastructures to accommodate tourists, and natural attractions offering high recreational value (Suwantoro, 1997, in Lubis, 2006).

Environmental Impacts of Ecotourism Activities: Ecotourism activities can have both positive and negative impacts on the environment and local culture. Negative impacts include pollution, degradation of landscapes and wildlife habitats, and poor management plans. Ecotourism activities, as defined by Hadinoto (1996), encompass: 1. Economic benefits: Ecotourism contributes to the economy through tourist expenditures, multiplier effects, and development linkages.
2. Synergism with the natural environment: Ecotourism

promotes the appreciation and understanding of ecosystems for tourists and local communities, emphasizing environmental preservation.

3. Incentives for sustainable resource management: Ecotourism motivates sustainable management of natural resources by governmental agencies, private sector, and individuals.

4. Public education: Ecotourism serves as a platform for educating the public about natural resources and their conservation.

Ecotourism development has significant impacts on the environment, encompassing physical, biological, economic, social, cultural, and political aspects. It induces ecosystem succession and leads to changes in traditional lifestyles and local community income. Ecotourism development activities have impacts on various aspects, including physical, biological, economic, social, cultural, and political elements. In terms of the natural environment, ecotourism development can influence ecosystem succession, potentially turning a natural ecosystem into a man-made one. Ecotourism activities in the Tondano Lake areas have had positive impacts. This include increased income for local communities, improved recreational facilities and accommodations for tourists, and the local people's creativity in making souvenirs from water hyacinth. This has also led to an increase in the population of water hyacinth due to its use in souvenir production. Additionally, tourism attractions such as fishing and water bikes have contributed to the area's appeal for tourists. The positive impact of tourism activities in terms of socioeconomic factors includes the increased income of local communities and the boost in rural employment. However, there are also negative impacts, such as ecosystem degradation due to the development of tourist attractions and potential social conflicts involving local communities (Lumintang, 1996). The changes in tourist activities impact nature in various ways, so it's crucial to prioritize sustainability. This involves understanding the natural environment, creating jobs, and involving local communities. The principles of sustainable tourism development are:

1. The environment has intrinsic value and can be a tourism asset. It should be utilized for the benefit of future generations.
2. Tourism should provide mutual benefits for local communities, the natural environment, and tourists.
3. The relationship between tourism and the natural environment must be managed sustainably.
4. Tourism activities must consider the scale of the natural and site properties.
5. There should be synchronization among the needs of tourists, the environment, and local communities.
6. Principles of adaptation to change should be implemented.
7. The tourism industry, local government, NGOs, and environmentalists must work together to implement these principles (Burn and Holden, 1997).

* * *

Definition of Ecotourism

Ecotourism generally deals with interaction with biotic components of the natural environment. Ecotourism focuses on socially responsible traveling, educating, creating awareness, personal growth, and environmental sustainability. The main idea of ecotourism is to create preserve and conserve the natural beauty and the cultural heritage of the tourist spots. It offers insight into the impact of human beings on the environment and appreciation of our natural habitats. Responsible ecotourism includes enhancing cultural integrity among local people, working towards spreading sustainable practices, and educating people to go for green business marketing. Ecotourism leads to positioning and promotion of recycling, water conservation, preserving energy efficiency, also protecting flora and fauna. In simple terms, ecotourism is the management of tourism and conservation of nature to maintain a balance between the requirements of "tourism and ecology" on one hand and community on the other hand.

❊ ❊ ❊

Definition of ecotourism by different authors and sources:

"Ecotourism is a nature-based tourism activity that contributes towards conservation through generating funds for protected areas, creating employment opportunities for local communities and providing them environmental education" (Boo, 1991).

"Nature-based tourism that is focused on the provision of learning opportunities while providing local and regional benefits, while demonstrating environmental, social, cultural, and economic sustainability" (Forestry Tasmania, 1994).

"Ecologically sustainable tourism in natural areas that interprets the local environment and cultures, furthers the tourists' understanding of them, fosters conservation and adds to the well-being of the local people"(Richardson, 1993: 1998).

"Nature-based tourism that involves education and interpretation of the natural environment and is managed to be ecologically sustainable. This definition recognizes that natural environment includes cultural components and that ecologically sustainable involves an appropriate return to the local communityand long-term conservation of the resource" (Australia Department of Tourism 1994: 2017).

"Travel to remote or natural areas which aims to enhance understanding and appreciation of the natural environment and cultural heritage, avoiding damage or deterioration of the "environment and the experience for others" (Figgis, 1993: 2008)

"Ecotourism is a form of tourism which fosters environmental principles, with an emphasis on visiting and observing natural areas" (Boyd & Butler, 1996).

"Ecotourism is a form of tourism inspired primarily by the natural history of an area, including its indigenous cultures. Ecotourism visits relatively undeveloped areas in the spirit of appreciation, participation, and sensitivity. The ecotourism practices anon-consumptive use of wildlife and natural resources and contributes to the visited area through labor or financial means aimed at directly benefiting the conservation of the site and the economic well-being of the residents" (Ziffer, 1989:1996).

Page -5

"**The International Ecotourism Society** (TIES, 2015) defines ecotourism as responsible travel to natural areas that conserves the environment and improves the well-being of local people".

United Nations World Tourism Organization (UNWTO), defines ecotourism as tourism that involves travelling to relatively undisturbed natural areas with the specified object of studying, admiring, and enjoying the scenery and its wild plants and animals as well as any cultural aspects past or present, found in these areas".

In 1996, the World Conservation Union (IUCN) defined ecotourism as environmentally responsible travel to natural areas, to enjoy and appreciate nature for the socioeconomic development of local people".

The United Nations Environmental Programme (UNEP) defines ecotourism as sustainable tourism, which follows a clear process that:

1. Ensures participation of stakeholders.

2. Ensures effective participation of the local community.

3. Acknowledge indigenous people communities' rights to say no to tourism development within the communities, land and territories.

4. Educating and making people aware of the critical issues faced by the tourist spots and their natural resources.

5. Promotes process for indigenous people and local communities.

6. To control and maintain their natural resources.

❉ ❉ ❉

"The Ecotourism Association of Australia defined ecotourism as nature-based tourism that involves education and interpretation of the natural environment and is managed to be ecologically sustainable".

"According to Tickell, ecotourism is travel to enjoy the world's amazing diversity of natural life and human culture without causing damage to either".

Ecotourism is an ecologically sustainable form of tourism that involves cultural understanding, appreciation, and conservation and also creates awareness of environmental impacts.

One of the major objectives of ecotourism is to improve the standard of living of poor people by eradicating hunger and poverty.

❉ ❉ ❉

PRINCIPLES OF ECOTOURISM:

According to the TIES ecotourism: - Ecotourism is about uniting and conserving with communities, for spreading awareness of sustainable travel. This means that those who implement, participate in and market ecotourism activities should adopt the following ecotourism principles:

Minimize physical, social, behavioural, and psychological impacts. Build environmental and cultural awareness and respect.

Provide positive experiences for both visitors and hosts. Provide direct financial benefits for conservation.

Generate financial benefits for both local people and private industry.
Deliver memorable interpretative experiences to visitors that help raise sensitivity to host
countries' political, environmental, and social climates.

Design, construct and operate low-impact facilities, which means not harming the
environment while constructing buildings or any other structures on the land or the water.

Recognize the rights and spiritual beliefs of the Indigenous People in your community and
work in partnership with them to create empowerment.

One of the very important principles of ecotourism is to make people in the society aware of the environmental problems faced by the tourist spots, common-pool resources, and wildlife habitats. The main idea of this principle is to position the relationship between human beings and the environment.

The very important principle of ecotourism is to ensure that people in the society do not break certain environmental laws, such as getting into protected areas without permission, overexploitation, or non-sustainable consumption of natural resources, and working towards a green economy.

Hetzer in the year of (1965) identified four pillars of responsible tourism which are defined below:

1. Minimizing environmental impacts

2. Respecting host cultures

3. Maximizing the benefits to local people and Maximizing tourist satisfaction.

❋ ❋ ❋

F ive sustainable development goals and their relationship with ecotourism:

1) LIFE BELOW WATER (SDG- 14)

2) LIFE ON LAND (SDG-15)

3) REDUCE HUNGER (SDG-2)

4) NO POVERTY (SDG-1)

5) CLIMATE ACTION (SDG-13)

Life below water (SDG-14) is one of the most important goals of sustainable development, 2030 which states the critical issues and problems faced by marine species due to the increasing number of human activities and irresponsible travelling. Oceans are the source of economic development, social development, and environmental protection. 71% of the earth is water which is an essen-tial natural resource for human beings, plants, and animals because without water there will be no life.

The relationship between life below water and eco-tourism:

Ecotourism helps in making tourists aware of the environmental problems faced by marine species and their natural resources.

Ecotourism creates an impact on the mind of the tourist to follow the guidelines and principles of the coastal tourist spots they are visiting. It also educates them to protect and conserve the common pool resources and coastal species that exist under the water.

Ecotourism can reduce the negative impact on mangroves, coral reefs, and wetlands found in the coastal areas, which are an important part of the marine ecosystem. These mangroves, coral reefs, and wetlands are threatened due to construction activities, dumping of garbage, and emission of harmful chemicals.

Ecotourism can help in generating employment opportunities for people who are living in the coastal areas and make them work towards marketing products that do not harm marine species. o It will end the idea of overfishing and will lead to sustainable fishing. It will also help in better sustainable practices for improving, managing, and restructuring coastal zones.

* * *

Definition of TourismTourism: The word tourism was first coined by Guyer Feuler in the year of 1905. The definition by Guyer Feuler (1905)defined tourism as "A collection of activities, services, and industries which deliver a travel experience comprising transportation, accommodation, eating and drinking establishments, retail shops and other hospitality services provided for individuals or groups traveling away from home (refer to the article written by Global Journal Management and Business Research).

* * *

Definition by Britannica: the act and process of spending time away from home in pursuit of recreation, relaxation, and pleasure, while making use of the commercial provision of services (Walton, J. K. (2024, July 7). tourism. Encyclopedia Britannica. https://www.britannica.com/topic/tourism).
Definition of Tourism by UNWTO (United Nations World Trade Organization): Tourism is a social, cultural and economic phenomenon which entails the movement of people to countries or places outside their usual environment for personal or business/professional purposes (https://www.unwto.org/glossary-tourism-terms).

The negative impacts of Tourism: have been defined as something that leads to the environmental degradation of air, water, land, and other types of natural resources. Air traveling is the least environment-friendly mode of transportation, as it contributes to global warming, which is caused due to emission of harmful greenhouse gases such as carbon dioxide and nitrogen oxides. This has caused harm to the marine ecosystem and the natural habitat of plants and threatened wildlife species within that tourist spots. It has led to climate change due to which there is a rise in the sea- temperature, also a change in precipitation, and unexpected weather events that will lead to more comprehensive problems such as soil, coastal erosion, depletion of the ozone layer, and melting of ice Glaciers.

Noise pollution caused by traveling through the air, roads, railways as well as recreational vehicles has led to annoyance, stress, and even hearing loss in humans, it harms and causes distress to the wildlife habitat, especially in sensitive areas. Water pollution caused due to ecotourism is due to the emission of untreated sewage, disposal of plastic waste into the water, emission of industrial chemicals, and effluents that harm the natural habitat of wildlife species, marine biodiversity, and its natural-common pool resources. Cruise ships are also one of the major sources of water pollution, as it burns a lot of fuel oil, the dirtiest fossil fuels available on the market which leads to the destruction of coral reefs and destroys the flora and fauna of the marine ecosystem. Water pollution reduces the oxygen level in the water, due to which the life below water (SDG-14) gets destroyed.

Page -1

Land degradation caused due to ecotourism includes the overexploitation of minerals, fossil fuels, fertile soil, deforestation, and harm to the wildlife habitat. This is because of the increased construction of tourism and recreational facilities that has created a direct impact on landscape natural resources. An increase in tourist facilities harmed land used for building accommodation, and the development of infrastructure through the use of building materials that are not sustainable. Solid waste and littering have also caused serious threats to the environment as tourist activities have led to the disposal of plastic waste and garbage, that harms the rivers, natural habitats, and roadsides. This degrades the physical appearance of the water and shoreline that causes the death of marine mammals. In mountain areas, trekking tourists generate a lot of waste, as they leave oxygen cylinders, campaign equipment, and garbage. Pollution of land, water, and air also lead to destroying the aesthetic natural environment. This is because of the illegal construction of resorts and hotels which harms and pollutes the natural habitat of land and marine species, and other types of common-pool resources.

❊ ❊ ❊

Summary on both positive and negative impacts of Ecotourism on Environment

Positive Impacts of Ecotourism:

- Reduces hunger and poverty by providing economic opportunities and income for local communities.

- Raises awareness about why it is important to protect vulnerable tourist spots and natural habitats.

- Educates people on the conservation of natural resources and the significance of protecting wildlife species.

- Promotes sustainable business practices and marketing, encouraging environmentally friendly tourism models.

- Generates employment opportunities, contributing to economic development and local livelihoods.

- Fosters cultural exchange and understanding, promoting respect for different traditions and communities.

- Provides financial benefits for conservation efforts, supporting habitat protection and species preservation.

- Empowers women by offering them equal opportunities and a source of income, enhancing their social and economic status.

- Helps combat the illegal trafficking of animal body parts by providing economic alternatives and increasing surveillance in protected areas.

- Minimizes human impact on marine and coastal environments through responsible tourism practices and marine conservation initiatives.

Negative Impacts of Ecotourism:

- Environmental degradation, including air pollution from increased carbon emissions, particularly in air travel, contributes to global warming and climate change.
- Climate change impacts, such as rising sea temperatures, altered precipitation patterns, and more frequent extreme weather events, leading to coastal erosion and ecosystem disruption.
- Noise pollution from various modes of transportation and recreational vehicles, disturbing both human and animal habitats, and leading to stress and hearing loss.
- Water pollution from untreated sewage, plastic waste, and industrial chemicals, damaging marine ecosystems, and threatening marine biodiversity.
- Destruction of coral reefs and marine life due to cruise ship emissions and the burning of fossil fuels, altering the delicate balance of aquatic environments.
- Land degradation through overexploitation of natural resources, deforestation, and the construction of tourism infrastructure, impacting wildlife habitats and ecosystem integrity.

- Solid waste and littering, particularly in remote areas, harming natural habitats, waterways, and the physical appearance of shorelines and landscapes.
- Disturbance and destruction of natural habitats due to illegal construction of resorts and hotels, impacting wildlife and local ecosystems.
- Aesthetic degradation of natural environments, including pollution and visual intrusion, detracting from the intrinsic beauty of tourist spots.

Some additional negative impacts to consider include:

- Overcrowding and overtourism: Popular ecotourism destinations may experience overcrowding, leading to further environmental degradation, increased pollution, and negative impacts on local communities.

- Cultural commodification: The commercialization of cultural practices and traditions for tourism can lead to the exploitation and distortion of authentic cultural heritage.

- Inequitable distribution of benefits: In some cases, the economic benefits of ecotourism may not reach local communities, leading to resentment and a lack of support for conservation efforts.

- Wildlife disruption: Ecotourism activities can disturb wildlife behaviour, feeding patterns, and migration, impacting the health and survival of species.

To mitigate these negative impacts, sustainable practices, responsible tourism models, and careful planning and management are essential. Collaboration between governments, local communities, tourism operators, and conservation organizations is key to ensuring that ecotourism has a positive net impact on the environment and local communities.

Ecotourism & Eco Friendly Clothes

Climate change is also one of the global problems which it has led to affecting the animals and plants which are already threatened. Loss of agricultural land led to poverty and hunger as land is the basic source of livelihood for poor people and even animals.

Ecotourism's relationship with the Life on land:

Educating the society to protect, conserve, and restore the cultural heritage of the particular natural habitat that exists in the life on land.

It helps in providing a better platform for creating awareness towards preventing poaching and illegal trafficking of animals' skin tissues and body parts.

Ecotourism can lead to reducing hunger and poverty by making better use of the land's natural resources, which ultimately leads to economic development.

It also makes the local and poor people aware of the environmental problems faced by the surroundings they are residing in.

Ecotourism can help in improving the standard of living and livelihood of poor people by making them work towards green marketing.

Ecotourism helps in the preparation and implementation of better guidelines to protect the animals, and plants in the forest which are critically endangered and are on the verge of getting extinct.

Reducing hunger and eradicating poverty (SDG-1 and 2) are some of the global issues and greatest challenges humanity faces. Many poor people are dying because of hunger, poverty, and lack of access to pure drinking water, sanitation, and hygiene. The two very important reasons for increasing hunger and poverty are:- droughts, conflicts and war, lack of adequate food supply and shortage, poor infrastructure, lack of job stability, unstable market, and gender inequalities.

Ecotourism and Agro-tourism relationship with hunger and poverty:

It helps in generating employment and leads to economic development.

Ecotourism can reduce gender inequalities which means creating equal access for both men and women farmers. Gender inequalities are one of the major barriers for women who want to work towards economic development and reduce hunger.

Ecotourism educates and creates awareness among both men and women farmers to work towards collaborative agricultural-based marketing.

Ecotourism leads farmers to work towards eco-organic farming and better sustainable practices. It also makes them protect and conserve their natural habitat.

Ecotourism & Eco Friendly Clothes

The trend of agro-tourism can help in making both men and women farmers more resilient so that they do not face any sought of discrimination.

Agro-tourism can provide farmers with a better platform to expand their business by attracting tourists.

Agro-tourism acts as a source of a steady income and helps in better innovation of food products and helps farmers to change according to the trend.

Agro-tourism helps farmers to do direct marketing which means it transforms a farmer into a price maker rather than a price taker. This helps them to directly interact with customers daily.

Climate Change is one of the biggest global concerns that has led to the destruction and degradation of the environment, through changes in climatic conditions. Climate change is caused due to emission of CFC (Chlorofluorocarbons), carbon dioxide and nitrous oxide (these gases are emitted due to the burning of coal and fossil fuels), methane (increasing number of livestock), and fluorinated gases (that are emitted by human activities). Tourism is also one of the reasons which contribute to climate change as many people travel through aircraft, cars, and cruises. Traveling by aircraft emits a lot of greenhouse gases that lead to depletion of the ozone layer and harms the air ecosystem also creates a lot of noise pollution that disturbs the aesthetic environment. Car causes a lot of pollution and contributes to climate change as it emits sulphur dioxide and greenhouse gases. Cruise harms the life below water due to which a lot of marine species and their natural resources get destroyed, as cruise generates tons of sewage and also emits toxic waste, which degrades the quality of water. This also leads to destroying the coral reefs, mangroves, and wetlands.

Ecotourism and Climate Action (SDG-13): have a huge relationship with each other as both go hand in hand. Ecotourism is the responsible way of traveling to make ourselves and others aware of environmental issues faced by the particular natural habitat and the wildlife species, whereas climate action can help in implementing better strategies, programs, and policies for reducing the impact of greenhouse gases, chlorofluorocarbons and other harmful gases in the natural environment. It can help in creating awareness towards protecting animals and plants listed in IUCN (International Union for Conservation of Nature).These creatures are critically endangered and vulnerable due to the impact of climate change on their natural habitat and common-pool resources. Ecotourism is the best platform to educate society to work in unity to prevent climate change by protecting marine, and land species, and conserving natural resources. The climate action program of Sustainable Development Goals is the best way of taking steps for working toward making people, students, and institutions aware of facts, and figures on how climate change can harm biodiversity. This program should also be implemented in rural areas as it is the right of every citizen to learn what environmental issues can be caused due to climate change.

Ecotourism & Eco Friendly Clothes

ECOCENTRISM

Ecocentrism is an ecological political philosophy that places high importance on living organisms, natural resources, and their natural habitat, rather than perceiving their value to human beings. Ecocentrism was first coined by the "Norwegian Philosopher Arne Naess" in 1973. Ecocentrism is also called deep ecology. Deep ecology is a broader concept that states that all the living environment and natural elements on the planet as a whole need to be respected by following the certain norm and moral values, of human beings. It describes the inter-relationship of nature with human beings and tries to strike a balance between both living elements.

For several years, human activities have created a lot of environmental issues by not respecting the wildlife, marine life, common-pool resources (land, water, air), and their natural habitat. The land is one of the vital parts of biodiversity as it is a home for millions of species, but due to the rise in human activities, it has been destroyed. The human intervention caused by humans leads to destruction that has led to deforestation, overgrazing, over-irrigation, mismanagement of agricultural land, overexploitation of land vegetative cover, (shrubs, trees, forest, grasslands), and also use of bio-industrial chemicals that cause harm to the land ecosystem. The land ecosystem is also getting overexploited because of hunting, poaching, and the illegal trafficking of animal tissues.

Water pollution is also one of the activities that cause a lot of harm to the marine ecosystem which constitutes coral reefs, wetlands, and other life forms that are critically endangered and endangered. Water is used by human beings for daily purposes such as cleaning utensils, washing the floors, and cars, drinking, and watering trees. Air is also polluted by human beings as they burn crackers and burn fossil fuels that lead to the emission of harmful gases such as carbon dioxide, nitrous oxide, and sulphur dioxide. These gases ultimately harm birds, destroy aesthetics, and result in noise pollution.

Some of the recent incidents and evidence of human activities causing environmental problems is where a pregnant elephant was fed pineapple that was implanted with firecrackers. Tragically, she suffered and died in Kerala inside the Silent Valley forest. Elephants' body parts, skin tissues, and tusks are also illegally trafficked for earning illicit profits. This shows human beings are human-centered, and that they have stopped caring for the environment. Recently, there was a tigress that had been killed brutally with sticks in the Phillbit Jari Village (Uttar Pradesh). The incident took place under a protected zone of the Pilibhit Tiger Reserve. Two juvenile leopards were also put to death inside the Katariniaghat wildlife division which is a protected area in Uttar Pradesh.

Marine mammals such as dolphins were used in the Second World War for detecting enemy mines, and navy ships, and sometimes they were used for implanting bombs to destroy the enemy ships which led to an outcry all over the world. Rhinoceros horns are illegally trafficked across the borders to make ornaments, jewellery, belt buckles, cups, buttons, and paperweights. Marine life trading is also one of the most criminal activities done coast to coast, few examples of these are: - crude oil, they are killed for their meat and blubber, and ribs are used for making art tools, utensils, and lamps.

✳ ✳ ✳

CHAPTER-2

INTRODUCTION TO ECO-FRIENDLY CLOTHES

WIDESPREAD ENVIRONMENTAL DEGRADATION AND THE CRUMBLING OF THE UNSUSTAINABLE FASHION INDUSTRY

BUSINESS MODELS have led us to examine and evaluate the clothes we wear. Consequently, the more ethical and discerning consumers amongst us are actively seeking out eco-friendly clothes to lower our carbon footprints when it comes to apparel buying.

'Eco-friendly' is often touted as a buzzword but what does it mean? The Sustainable Technology Education Project (STEP) defines eco-friendly fashion as clothes made from organic fabrics, without dangerous chemicals like dyes, and originating from fair trade practices. It also includes clothes manufactured from materials that are recycled and reused.

It is made from organic fabrics: To begin with, let us divide cloth fibers into natural fibers that are biodegradable and synthetic fibers that are non-biodegradable (Utter Dahl, 2020). Natural fibers are sourced from plants and include cotton, hemp, linen, and bamboo. On the other hand, natural fibers derived from animals are wool, silk, cashmere, and angora amongst others.

Starting with plant fibers, cotton is known for being biodegradable and comfortable to be worn next to the skin. However, cultivating this crop calls for a tremendous amount of water and land. Additionally, conventional cotton farmers employ toxic fertilizers which have widespread harmful effects. To alleviate this, organic cotton is on the rise as a renewable fiber hopefully replacing conventional cotton in the long run.

Hemp has the potential to be a strong and sustainable fiber since it does not require much water and nutrition, eliminating the need for pesticides. Interestingly, growing hemp prevents soil erosion as its long roots bind the soil. Sadly, there are not many countries that legalize the cultivation of this formidable plant.

Linen is sourced from the flax plant and requires less water and no fertilizer or pesticide. If you have ever worn linen clothes before, you would be familiar with how stiff and strong this fabric is. Very popular in warmer climates, linen is far more breathable and cooler than cotton.

Bamboo is growing in popularity as a renewable fiber because it can grow easily and rapidly without the need for pesticides. Unfortunately, converting the fibers of bamboo into usable textiles involves using hazardous chemicals.

Turning to animal fibers, wool from sheep is breathable, durable, and absorbent. From an industrial point of view, it is also an efficient use of by-products from the livestock animal industry. Nonetheless, vegan consumers who refrain from any use of animal products will shun buying woolen clothing. Also, there is the controversial practice of 'molesting' in Australia which causes insurmountable suffering to merino sheep. This can prove to be a major deterrent to ethical shoppers who will demand cruelty-free fabrics.

* * *

Ecotourism & Eco Friendly Clothes

Silk first came from China and is made by silkworms. Silk is soft and comfortable on the skin. As with other animal fibers, it involves cruelty as the worms are boiled, making it a definite No for ethical shoppers. To cater to a growing vegan customer segment, companies are researching vegan silk as an alternative to worm silk.

Cashmere feels luxuriously soft on the body and is marketed as a luxe winter fabric. Cashmere is taken from the cashmere goats in certain Asian regions for instance Iran, Pakistan, and Afghanistan. Alongside the cruel aspect present, the goats also degrade the soil unintentionally through their extensive grazing.

Angora is rabbit fur and known for being immensely soft. To maximize profit and increase production efficiency, rabbits' fur is ripped off mercilessly. Naturally, there is unspeakable pain and animal welfare is neglected. To make matters worse, rabbits are prey animals that are extremely nervous and timid.

Manufactured without dangerous chemicals: The process of dying fabrics impacts negatively on the natural marine environment. Unscrupulous factories that are not regulated by the government leak untreated toxic waste such as formaldehyde and dioxins into the waterways (Livingston, 2020). As a result, water pollution occurs. Consequently, aquatic life and human beings who reside around these industrial sites are unfairly and adversely affected in turn.

In comparison, eco-friendly clothing reduces water pollution in its tracks as ethical clothes producers strive to select greener, renewable fibers. Not only that, but the producers also monitor the type of dye to be used by selecting either natural dyes derived from plant and mineral sources or low-impact dyes to minimize the environmental impacts. What is more, some producers go as far as offering uncolored fabric which is well suited to people with sensitive skin.

Employs fair-trade practices: Many of us read with sadness the news of the 2013 factory incident in Bangladesh, which took the lives of 1134 unfortunate employees (Bick, Halsey & C. Ekenga, 2018). Admittedly, this is only one of the incidents out of the many that occur worldwide endangering the health and safety of workers. In other words, the old capitalist business model of free trade is no longer the optimal one and a more equitable fair-trade model can be witnessed around us. For this reason, the World Fair Trade Organization (WFTO) aims to put people and the planet above business profits and its members adhere to ten trade principles.

According to Hodakel (2020), fair trade clothing is clothing manufactured based on ethical trade standards. There is a movement sweeping across the globe where consumers are demanding increased transparency to vote with their dollars. If so, how will clothing producers cater to the demands of this growing market segment? For a start, eco-friendly clothes producers could apply to be fair-trade certified.

Besides, they can evaluate their current supply chain practices to adhere to better standards. What can you do as a consumer? By researching and opting to shop with fair-trade certified brands, we can avoid being 'green washed' by unscrupulous companies.

Ecotourism & Eco Friendly Clothes

In brief, we now know that eco-friendly clothes are ideally sourced

Organic fabrics are made without dangerous chemicals, manufactured in line with fair trade practices, and close the loop by using recycled or reused materials.

Despite there being some drawbacks associated with certain natural fibers, using synthetic fabrics still poses far greater complications in terms of environmental and health hazards. To eradicate the suffering of animals used for their fur, there may come a time when animal fur can be successfully cultivated in labs and used to produce textiles. Furthermore, initiatives such as the Better Cotton Initiative (BCI) are already in place considering the impact of cotton production on the environment.

In the meantime, scientists continue to research plants in hopes of uncovering new dyes which are environmentally friendly to phase out chemical dyes. Finally, numerous companies are moving in the right direction by transforming their business model into one that is socially and environmentally responsible.

More about eco-friendly clothes and how they can prevent animal trafficking

Eco-friendly clothes are defined as "clothes which are made up of pure cotton and woolen fabric, that are biodegradable and decomposable". We are well aware of what eco-friendly products are and how it helps in protecting the environment, but still, we do not use them in our daily life. This concept of Eco-friendly clothes is not followed everywhere around the world, as nowadays more and more people are only wearing clothes that are made up of leather, polyester and nylon, and acrylic. These fabrics are not eco-friendly and neither they are sustainable. Eco-friendly clothes are eco-smart that work towards better sustainable practices and reduce the negative impact on the environment. By mentioning sustainable practices it means that eco-friendly clothes help in generating employment opportunities for poor people that ultimately help in reducing hunger and poverty.

Eco-friendly clothes help in keeping our bodies cool and protect us from diseases caused by ultraviolet radiation. These clothes are printed with eco inks and beautiful messages that can help in spreading awareness among the society to protect wildlife and its natural habitat. Ecofriendly clothes conserve the natural resources of tourist spots by educating people not to pluck flowers and throwing plastic waste in the tourist spots which can lead to the destruction of flora and fauna. It also prevents the exploitation of animals because many people from around the world traveling to the tourist spots want to click pictures but do not want to go deep into why those animals have been conserved and why is it protected in their natural habitat. It also makes society aware of their responsibilities and duties towards protecting nature. It reduces the carbon footprint which means it consumes less energy and does not emit harmful gases. Ecofriendly clothes help in promoting the word Ecotourism which can help in educating people about responsible and sustainable tourism. Eco-friendly clothes work toward preventing animal trafficking which is one of the serious and offensive crimes in the field of the environment.

* * *

What does animal trafficking means: it is defined as the smuggling, trading, distributing, and manufacturing of animal body parts and using their tissues and skins for business purposes. These activities are done across the borders to earn more profits, by hunting, poaching, and killing animals some of which are also listed under the IUCN (International Union for

Conservation of Nature) red list (this will further be discussed in the book). Products made out of animal body parts, which are illegally traded across the borders and are highly in demand constitute exotic pets that are defined as the unusual animal kept within human households which is relatively unusual to keep or is generally thought of as wild species rather than as a pet. Some of the products which are made up of these exotic pets are sweaters, jackets, scarves, handbags, shoes, food, traditional medicine, clothing, and jewellery made up of tusks, fins, skins, and shells, horns, and internal organs.

Some of the few examples of land species that are being killed for clothing and business purposes are the following: Silkworms, Rabbits, Seals, Foxes, Sheep, Lambs, Chiru, Minks, Beavers, Dogs and Cats, Bears, Tigers, and Rhinoceros. Examples: leather and fur produced out of animal skins are main contributors to agricultural business as they help in making profit through dairy products and their flesh is created through toxic tanning of animal skin. Today most of the clothes are made out of cattle skin such as lamb and deer skin is used for producing soft leather which is more expensive apparel. Deer and elk skin are widely used for making work gloves and indoor shoes. Pigskin is used for producing wallets and seats of saddles. Buffalo, horses, goats, alligators, crocodiles, dogs, snakes, ostriches, kangaroos, oxen, and yaks are also used for leather. Kangaroo skin is used to make items that should be strong and flexible—it is the material most commonly used in bullwhips. Kangaroo leather is purchased and favoured by the motorcyclist as it helps them to make better tracksuits and gloves which leads to Kangaroos depletion. Also, Kangaroo skin helps in producing soccer footwear.

Ostriches used to be hunted in the 19th century as they were very famous for meat and leather. Ostrich leather is currently used by many major fashion houses such as Hermes, Prada, Gucci, and Louis Vuitton to make wallets, clothes, seats, shoes, and jackets. Rhinoceros is also one of the critically endangered animals which are being killed to make ornaments, jewellery, cups, belt buckles, hair pins, and paperweights. It is also said that Rhinoceros are killed because their horns can cure diseases such as typhoid, vomiting, headaches, demon possession and snakebites, etc. This is known as "Aphrodisiac". However, these are myths due to these reasons many Rhinoceros get poached. Tigers are also poached for their skins which help in making clothes and are used in sculptures. So, we should convert and shift ourselves towards eco-friendly clothing as it can save millions of known critically vulnerable and endangered species as depicted above. We should go for slow fashion rather going for fast fashion, as slow fashion can lead to an increase in the level of climate change, due to the emission of harmful gases. The production of unsustainable clothes leads to create a lot of air, water, and land pollution.

Ecotourism & Eco Friendly Clothes

IMPORTANCE OF ECO-FRIENDLY CLOTHES

MAKING THE TRANSITION TO ECO-FRIENDLY CLOTHES is imperative to restoring social, environmental, and ethical justice. The previous section described how numerous positive attributes of eco-wear help protect our environment. Now, we will discuss the importance of eco-friendly clothes to consumers, businesses, and regulators.

Eco-friendly apparel has the exceptional advantage of being safe to wear on our skin, especially organic fabrics. This is because these fabrics are holistically derived from nature, right from cultivating a non-genetically modified textile seed, colouring the fabric in the least harmful process, and the final step of composting at the end of the fabric's life. Why should it matter to us as end-users? Wearing synthetic or toxic textiles can be detrimental in the long run as our skin is highly permeable. As a result, these substances can build up in our bodies, leading to respiratory ailments and cancer, amongst others.

Eco-friendly wear can also meet the inevitable soaring demand for clothes as the global population continues to rise alarmingly to an estimated 8.5 billion people by 2030 (Lensing, as cited in Pulse of the Fashion Industry, 2017). Consequently, fashion businesses will face severe scarcity issues in the areas of land, energy, water, and human resources required as they are being depleted by human activities. Eco-friendly clothes can address these operational scarcity issues by tweaking the industry's current business model to a more sustainable one over the long term. This can be accomplished as eco fibers visibly minimize strain on natural resources by being biodegradable, of optimal quality, and embracing an all-inclusive business outlook.

Eco-friendly wear is ideally composed of sustainable fibers that are biodegradable thereby closing the loop at the end of the fabric's life. This decreases energy, water, chemical, land, and waste footprints on our planet caused by unsustainable fashion industry practices. Having said that, transforming and regenerating the fashion industry will realistically require regulators to take a more proactive stance in formulating legal frameworks.

As an example, regulators can offer incentives or tax breaks for progressive companies that harness renewable energy. Not only that, but governments can also be more stringent on companies' water and chemical usage. A uniform global industry auditing body can set the benchmarks for the fashion industry to conform to (Pulse of the Fashion Industry, 2017, p. 106 107).

In conclusion, eco-friendly clothes are important to consumers for wearer safety and health, to corporations for addressing resource scarcity, and to regulators for driving the change forward and rewarding sustainable industry players.

Businesses can profit by recognizing the upside and potential of untapped value lying hidden in sustainable value chain practices. The segment with the largest impact would be consumers themselves and greater awareness, education, and grassroots efforts are needed to achieve this aim. Nonetheless, with the collaboration of consumers, companies, and the government, we may witness the much-needed change taking root in the fashion industry.

❊ ❊ ❊

Ecotourism & Eco Friendly Clothes

SUSTAINABLE FASHION: MEANING AND IMPORTANCE

AS SUSTAINABLE FASHION IS HIGHLY DISCUSSED AND IS REVOLUTIONIZING THE FASHION INDUSTRY, deciphering it beyond the formal definition is key to comprehending the movement that is taking place across the industry. This section will highlight the meaning and importance of sustainable fashion as an alternative to the current business model which is highly unsustainable. Green Strategy presents sustainable fashion as "partly about producing clothes, shoes, and accessories in environmentally and socio-economically sustainable manners, but also more about sustainable patterns of consumption and use, which necessitate shifts in individual attitudes and behaviour" (Brismar, 2020, para. 3).

From a consumer and manufacturer's perspective, there are seven forms of sustainable fashion identified on Green Strategy, being "on-demand and custom made; green and clean; highquality and timeless design; fair and ethical; repair, redesign and upcycled; rent, lease, and swap; second-hand and vintage" (Brismar, 2019, para. 3). Now, we will proceed to look at some of the advantages of sustainable fashion.

First, from a social point of view, sustainable fashion is here to overhaul unfair labor practices currently in place worldwide. For instance, the report Pulse of the Fashion Industry (2017) identifies labor issues such as minimum wage vs. living wage, gender pay gap discrepancies, treatment of workers, and child labor. These issues impact prevailing social conditions, especially in Low- and Middle-Income Countries (LMIC). Sustainable fashion reduces and eventually eliminates unfair working practices as it upholds principles of being fair and ethical in operations.

One of the clothing brand Levi's implemented its Worker Well-Being Initiative program in 2011. This move is to be lauded as other industry players will note the trend and awareness of increasing social responsibility.

Next, environmentally, sustainable fashion seeks to reduce energy footprints over the long term. Businesses will increasingly come face to face with water scarcity, resource depletion, and global warming if we carry on plundering our planet unconscionably. Disruptive solutions are critically needed to mitigate the decades of damages caused by unsustainable fashion. Sustainable fashion strives to close the recycling loop, offer renewable materials mixes, use green energy, and zero dangerous chemicals. An inspiring step in this direction is Adidas' No Dye design principle which aims to avoid water or chemical use in dyeing (Pulse of the Fashion Industry, 2017).

Finally, sustainable fashion has the potential to steer businesses into new frontiers. As competition increases in the capitalist model, natural resources will continuously deplete ultimately leading to extreme costs in business operations. To thrive and sustain growth and profitability, fashion brands need to zero in on the fair-trade element present in the philosophy of sustainable fashion. Annually, €160 billion lies untapped in the form of value opportunity of sustainable fashion (Pulse of the Fashion Industry, 2017). Environmentally, these comprise a reduction in water usage, energy emissions, waste produced, and worker illnesses. Whereas socially, the values are in reduced injuries at work and increased community engagement.

✳ ✳ ✳

Ecotourism & Eco Friendly Clothes

To sum up, sustainable fashion beckons all of us to scale our footprint on the earth and resume socially, environmentally, and financially sustainable modes of doing business. As human and animal populations are projected to climb upward outpacing the supply of natural resources available, businesses and consumers alike will come under pressure. At the same time, it is in our hands and hearts to leap forward in embracing a better way of consuming and producing clothing, if we desire to leave the earth a better place for the future generation to come.

CONCLUSION:

Sustainability is defined as the consumption of natural resources in such a way that the present and future generation can meet their basic needs and wants.

Sustainability is one of the major problems in all parts of the country, as hardly anyone is aware of why there is a need to promote the same (sustainability). Sustainability can help in creating a better future and better tomorrow through using proper conservation and planning techniques. The above statements made about sustainable fashion have been written for creating awareness towards the betterment of the people and society. Nowadays, people are going for fast fashion which is an unsustainable form of making a better choice while purchasing clothes. We spend far too much money on fashion for improving our personalities and styles, but we forget due to this we are putting lives at stake. The trend of fast fashion leads to animal trafficking which is one of the criminal offenses in our society and all parts of the world. Animal trafficking is caused due to lack of knowledge about the environmental problems, no awareness among the people in the society, lack of support from the top management, orthodox views about animal body parts and skin tissues, greediness for money, and lastly, improper implementation strategies and programs for the conservation of animals and various other species in the planet.

HOW ECO-FRIENDLY CLOTHES CAN PREVENT ANIMAL TRAFFICKING AND PROMOTE ECOTOURISM:

ANIMAL TRAFFICKING IS A BILLION DOLLAR INDUSTRY and an organized crime that threatens to decimate endangered species. Not only fashion has exploited animals and contributed to animal poaching but it has also wrecked irreversible harm to the environment. For these reasons and more, we will now closely examine first, how eco-friendly clothes can prevent animal trafficking and then, how eco-friendly clothes promote ecotourism.

One, eco-friendly clothes are made from natural fibers which include both animal and plant fibers. Increasingly these clothing are being made by sourcing Earth-friendly plant fibers such as hemp, bamboo, cork, mushroom, and fruit waste as it is more readily available and does not need years to grow as in the case of animals. Therefore, no animals are trafficked for their skins or fur to be used as material. Furthermore, there are no financial incentives for animal poachers to remain in the illicit animal trafficking business.

Second, eco-friendly clothes are made with fair-trade practices that alleviate environmental and social damages that may occur. An exemplar business would be Eileen Fisher which has ethical supply chains and uses handmade clothes from undyed alpaca fibers in Peru (Stein & Altmann, 2015). Undyed fibers ensure almost no chemical usage which improves safety for workers and no hazardous substances being leaked into waterways.

Page -35

CASE STUDY ON FUTURE PROSPECT OF ECOTOURISM IN RAMDHURA

Kalimpong

Important facts

Aim of the case study is to study:

The present scenario of Ecotourism in India.

Also, to see the future prospect of Ecotourism

To, explore and review sustainable tourism practices for creating better future of our Earth.

Introduction to Ramdhura Kalimpong

Ramdhura is a small village in Darjeeling, West Bengal, covering an area of approximately

1,053.60 km2 (406.8 sq. mi). Its geographical coordinates are 27°9'41" N longitude and 88°33'48" E latitude, with an elevation ranging from 5000 to 5500 feet and 15 Km away from (DeloHill comes under the state of West Bengal, it is situated at an altitude of 1704 metres (5590 ft). Delo is the highest point of Kalimpong town. Ramdhura village is situated 17 km away from Kalimpong town.The village is situated among pine forests and other trees. It is famous for its homestays, located in the mountains and facing Kanchenjunga. Additionally, the village offers a breathtaking view of the sunrise.Ramdhura is home to various attractions such as the picturesque Teesta River that flows through a verdant valley, serene pine forests, and the vibrant Cinchona Plantation (The plantation, started by the British in 1901, is part of Burmaik Cinchona Plantation, which is under the jurisdiction of Munsong Cinchona Plantation in Kalimpongand consists of 96 families).During your visit, you can see a variety of colourful birds and butterflies. The name "Ramdhura" comes from the local language, where "Ram" means a Hindu god and "Dhura" means village.Tourist spots that one can travelto in Ramdhura Kalimpong are as below: Ramdhura has several stunning viewpoints, such as:

Hanuman Tok and Damsang Fort, which offer breathtaking views of valleys and mountains.

Delo View Point is a magnificent park and a popular destination for paragliding.

Tinchuley Viewpoint is another delightful spot to visit. Other recommended tourist attractions near Ramdhura include Durpin Dara Hill, where visitors can enjoy panoramic views of the Chola Range of Sikkim.

Echey Gaon is a picturesque Gorkha village renowned for its organic farming methods.

The Bhutia Monastery, located just 1 KM away from Ramdhura Village, is also worth a visit. Trekking: You can trek through the forest to the Cinchona plantation, which is a haven for birdwatchers.

Paragliding: You can paraglide from Delo.

White River Rafting: You can travel to Teesta Bazaar to experience white water rafting.

The Darjeeling Zoo, also known as Padmaja Naidu Himalayan Zoological Park, is a 67.56-acre wildlife sanctuary located in Darjeeling, West Bengal. Established by the Department of

Education of the Government of West Bengal in 1957, its primary aim is to research and safeguard the Himalayan fauna. The zoo is renowned worldwide for its conservation breeding programs of endangered species such as Tibetan Wolves, Snow Leopards, and Red Pandas.
There are about 156 animals found in this Zoological Park.

Future Prospects of Ecotourism in the Ramdhura village

a) The village offers a plethora of thrilling activities, including rock climbing, para-gliding, mountain biking, hiking, trekking, and camping. It has a noble vision of creating an eco-park to protect the local flora and fauna, as well as a heritage centre to celebrate its culture.

b) To support the local community, it's essential to allocate funds for the 40% of villagers who are either unemployed or work outside the village. The remaining 60% work as plantation workers and earn a meagre salary of Rs. 4000-5000 per month. While many villagers are interested in ecotourism, they face financial obstacles in starting their businesses.

c) With the government and Cinchona department's backing, ecotourism could flourish. Effective advertising, tourism information centres, public transport improvements, and hospitality training programs are essential to attract both local and foreign visitors.

d) Community involvement is critical in minimizing conflicts between tourists and locals, as only 20% of villagers are currently involved in ecotourism. Training programs will provide residents with the necessary skills to accommodate tourists.

e) The village confronts safety concerns and inadequate infrastructure, which can be addressed by installing a water reservoir to ensure a reliable water source for both villagers and the tourism industry. The village offers a plethora of thrilling activities, including rock climbing, para-gliding, mountain biking, hiking, trekking, and camping. It has a noble vision of creating an eco-park to protect the local flora and fauna, as well as a heritage centre to celebrate its culture.

f) To support the local community, it's essential to allocate funds for the 40% of villagers who are either unemployed or work outside the village. The remaining 60% work as plantation workers and earn a meagre salary of Rs. 4000-5000 per month. While many villagers are interested in ecotourism, they face financial obstacles in starting their businesses.

g) With the government and Cinchona department's backing, ecotourism could flourish. Effective advertising, tourism information centres, public transport improvements, and hospitality training programs are essential to attract both local and foreign visitors.

h) Community involvement is critical in minimizing conflicts between tourists and locals, as only 20% of villagers are currently involved in ecotourism. Training programs will provide residents with the necessary skills to accommodate tourists.

i) The village confronts safety concerns and inadequate infrastructure, which can be addressed by installing a water reservoir to ensure a reliable water source for both villagers and the tourism industry.

LAKSHYA MEHTA

* * *

CASE STUDY ON SOCIAL IMPACTS OF ECOTOURISMIN INDIA

Important facts

Aim of the case study is to study the impact of ecotourism on the Indian economy and suggest policy recommendations.

Introduction to Indian Ecotourism

The tourism industry is rapidly expanding and is currently the world's largest industry. India's tourism is closely related to its rich cultural heritage and traditions. Ecotourism has gained global attention, especially in developing countries, as it is linked to sustainable development, conservation efforts, and community development strategies. Its goal is to promote responsible travel to natural areas that preserve the environment and improve the well-being of local people. However, the rapid growth of tourism has negatively impacted the environment and socio-cultural concerns. This study analyses the challenges and opportunities of eco-tourism and evaluates whether it is a viable option for sustainable development and conservation of India's rich culture and environment.

Tourism in India is a vital economic sector that is expanding rapidly. In the last two decades, India has opened its doors to international visitors to increase foreign earnings and boost its economy. In 2015, the country had 8.03 million international tourists, a 10.2% increase from the previous year. However, India's tourism infrastructure is struggling to keep pace with the industry's growth, and there are evident issues in the accommodation, transport, and personnel sectors.

Ecotourism is a sustainable form of tourism that involves visiting fragile, pristine, and relatively undisturbed natural areas with a low-impact and small-scale approach. Its purpose may be to educate the traveller, provide funds for ecological conservation, directly benefit the local community's economic development and political empowerment, or foster respect for different cultures and human rights. Since the 1980s, environmentalists have considered ecotourism a critical endeavour to preserve destinations untouched by human intervention for future generations. It advocates preventing irreversible changes to environmental assets, loss of ozone layers and living species, and damage to essential ecosystem functions such as tropical and primary forests, wetlands, and coral reefs. The same principle applies to human resources, local cultures, traditions, livelihoods, and the land on which they are based, which need to be respected.

India offers a unique form of sustainable tourism called ecotourism that revolves around exploring natural locations including wildlife sanctuaries, forests, and tea and spice plantations. This type of tourism helps preserve natural habitats while also contributing to the socioeconomic development of local communities.

If you're interested in experiencing ecotourism in India, there are several top destinations to consider such as Honey Hills - Thenmala Eco-Tourism in Kerala, Bandipur National Park in Karnataka, Sundarbans National Park in West Bengal, Chilika in Orissa, Kaziranga National Park in Assam, and Ranthambore National Park in Rajasthan.

Ecotourism & Eco Friendly Clothes

In addition to these locations, several states in India are actively promoting ecotourism, including Ladakh, Himachal Pradesh, Uttarakhand, Delhi, Meghalaya, Arunachal Pradesh, and Assam.

To protect and conserve wildlife, India currently has around 80 national parks and 441 sanctuaries. In fact, Thenmala was the first planned ecotourism destination in India and was selected by the World Tourism Organization as a premier eco-friendly project.

In the 1970s and 1980s, India saw the development of ecotourism. The first ecotourism destination, Thenmala, was planned in Kerala. Ecotourism involves visiting natural areas to support conservation efforts, while causing minimal impact and observing flora and fauna in their natural environment. The concept of ecotourism dates back to the 1970s, and in 1990, the International Ecotourism Society defined it as responsible travel to natural areas that conserves the environment and enhances the well-being of local people.

The ecotourism market in India had a value of $2.24 billion in 2019 and is expected to grow to $4.55 billion by 2027, with a compound annual growth rate of 15.7%. Additionally, the sustainable tourism market in India is expected to be worth $26.01 million in 2022 and projected to reach $151.88 million by 2032, with a CAGR of 19.3% during the period of 2022 to 2032.

Ecotourism is a sustainable mode of travel that prioritizes the protection and preservation of the natural environment. The practice entails responsible travel to natural destinations to safeguard the natural habitat and improve the quality of life for local inhabitants. India offers numerous ecotourism destinations, such as the backwaters of Kerala, the rainforests of Karnataka, the villages of Andhra Pradesh, the snow-capped peaks of Uttarakhand, the desert-scapes of Rajasthan, the pristine beaches of Goa, and the meadows and riverside trails of Sikkim. Kerala, in particular, is an excellent starting point for ecotourism since it was the first state to implement planned ecotourism and is situated in the Western Ghats, one of the world's top 18 biodiversity hotspots. You can find many endangered species of wildlife here. Other states with renowned ecotourism offerings include Arunachal Pradesh, with its 26 remote mountain valleys and indigenous tribes; West Bengal, which boasts coasts, hills, tea gardens, and forests; Sundarbans National Park, a UNESCO World Heritage Site; Jaldapara National Park, home to Indian onehorned rhinoceroses; and Neora Valley National Park, known for its diverse flora and fauna.

* * *

Challenges related to Ecotourism in India: In India, ecotourism is facing various challenges that need to be addressed. These challenges include a lack of proper infrastructure, such as transportation and sanitation facilities, insufficient marketing support from the government, inadequate planning and organization, deforestation, pollution, disposal of campsite sewage in rivers, lack of interest from people, and a lack of cooperation from the masses.

Ecotourism has the potential to benefit both protected areas and local communities by connecting biodiversity conservation with local, social, and economic development. However, economic development often comes at the expense of the environment. For instance, natural resources are destroyed to create souvenirs, and overused tracks can lead to soil erosion and damage to vegetation. Other challenges in Ecotourism sector in India includes the following:

Destruction of wildlife and natural habitat of tourist spots.

Contributes to Global Warming

Throwing of garbage on the roadside that affects the life below water and life on land.

It also creates a market for prostitution and selling of drugs.

Commoditization of cultural tourist spot into a business spot. (What is commoditization:

Commodification in tourism is the use of a religious spots and artifacts to make a profit.

* * *

This can contain comprising intangible cultural heritage into business names, branding, logos, and products.

Overuse- exploitation of ocean natural resources, which are limited and needs to be conserved.

Lack of support from the top management for improving the rules and regulations of tourist spots in India.

Travelers should not carry food and beverages inside the tourist spots, and they need to read every instruction before entering the boundaries of natural areas.

Nowadays, technology has advanced over the period of time, so as the society, so they should avoid taking devices that causes noise pollution, andled to destruction of flora and fauna for example: Radios, tape recorders, and other electronic entertainment, equipment and VR goggles.

Policy to be Implemented to improve Ecotourism in India

Preservation of cultural and natural heritage

There should be boards and signs for not harming both land and water resources.

Encourage the tourist to buy eco-friendly products.

Developing strategies for effective sustainable tourismtokeep the environment clean and green.

Good information, research and communication is needed for providing systematic information of thetourist spot such as: history, principles and norms.

Private sector should come forward to contribute towards

economic development through promoting better employment opportunities for poor people.

Agencies and other organisations should come together for creating sustainable business model that can benefit local and poor communities.

Recruiting staff should be done by setting proper standards in the tourism and hospitality sector.

Integrating social and economic planning for commencement of any major projects.

Infrastructure development is very important to improve transportation means and methods.

✽ ✽ ✽

A STRATEGIC APPROACH FOR SUSTAINABLE CLOTHING IN INDIA TOWARDS SLOW FASHION

Aim of the case study is to provide guidance to accept slow fashion using two planning tools namely Framework of Strategic Sustainable Development and Leverage point system designed by Donella Meadows.

The first approach Framework of strategic Sustainable Development includes 5 steps that are as following:

System
Success
Strategic Guidelines
Action level
Tool level

In the first, step, authors recognize the concept of slow fashion within the broader fashion industry and its impact on society and the environment. Slow fashion involves a group of manufacturers, retailers, brands, and designers who prioritize sustainability in their practices. This system is a smaller part of the larger fashion industry, which is in turn a subset of society, and ultimately reliant on the ecosphere for its existence. Therefore, each component of the fashion system is dependent on the environment for its sustainability.

Second step, in this step, we establish the parameters for success in planning for slow fashion. These parameters are the core principles necessary for a sustainable society. Isabella Oriani provides a generalized sustainable system from which these principles are derived, as outlined below. A sustainable society should not subject nature to systematic increases in: 1) concentrations of substances extracted from the earth's crust, 2) concentrations of substances produced by society, 3) degradation by physical means. Additionally, people in such a society should not be subject to conditions that systematically undermine their ability to meet their basic needs. Slow fashion involves producing high-quality, timeless clothing that can be worn across multiple seasons, and is a business strategy that prioritizes sustainability. In contrast, fast fashion is a business strategy that prioritizes quick response to consumer demand through an efficient supply chain.

Third step, Various stakeholders such as farmers, designers, manufacturers, and retailers have implemented diverse sustainability initiatives. These include growing organic cotton, wool, and bamboo fibres. However, reaching a consensus on the desired future scenario is challenging due to differing perspectives and vested interests. To address this issue, the authors propose a flexible guideline system called 'back casting' for the fashion industry. Back casting involves envisioning a desired future and planning how to achieve it from the current situation. Slow fashion initiatives must meet three criteria: 1) align with the established vision, 2) provide a flexible platform for future improvement, and 3) produce a sufficient return on investment. By adopting this approach, the fashion industry can promote sustainability while maintaining flexibility.

Fourth step, to achieve our vision in slow fashion, we need to take tangible actions at the Actions level. Here are some identified initiatives: 1. We can increase the lifespan of textile materials through recycling and redesigning. 2. We can re-use used materials. 3. We should use organic materials such as organic cotton and vegetable dyes. 4. We should implement environmentally friendly production processes.

Fifth step, within our Tools category, we have identified valuable resources that can enhance comprehension of the system and streamline retrospection. To address this matter, our writers have proposed the creation of a standardized eco-label that can be utilized throughout the entire lifecycle of clothing - from the inception of fibres to disposal. It is imperative that these eco-labels are accessible to all and affordable for small and medium-sized businesses.

Presently, there exists a lack of uniformity in the definition and approval process of various ecolabels such as Global Organic Cotton Standard (GOTS), Eco-Tex, ISO-14000, and others. The Second is 12-point leverage system designed by system-thinker Donella Meadows in Slow fashion:

Donella Meadow's 12-point leverage point system is utilized to analyse the fashion industry and identify the "levers and drivers" within the system where a small shift can cause a significant change in the ecosystem. Below are the important findings related to the above points:

3.1 Structure of Information Flows
The information flow in the fashion industry can be categorized into two types: from business to consumer and from business to business.

3.1.1 Business-to-Consumer Information Flow
Previous studies have indicated that consumers may be willing

to pay higher prices for "green" products, but it is challenging to communicate the Eco-friendliness of a product to the consumer. Lack of awareness among consumers about environmentally friendly clothing is hindering consumption. Consumers also have a simplistic view of the "greenness" of products. For instance, when competing for the "greenness" of cotton manufacturing versus polyester, the average consumer might assume that cotton is more environmentally responsible because it is a natural fiber, which is not necessarily true. One way for manufacturers and retailers to communicate the Eco-friendliness of products to the consumer is through Eco-labels.

3.1.2 Structure of Information Flow within the Textile Supply Chain

Within the textile and fashion industry, the non-availability of environmentally friendly raw materials is identified as a key constraint in using Eco-friendly raw materials. For example, while India is the world's largest producer of organic cotton, the majority of organic cotton is exported, and it is not easily available locally. Organic raw materials are supplied by a limited number of suppliers and farmers, which restricts the bargaining power of larger retailers.

Ecotourism & Eco Friendly Clothes

Fourth step, to achieve our vision in slow fashion, we need to take tangible actions at the Actions level. Here are some identified initiatives: 1. We can increase the lifespan of textile materials through recycling and redesigning. 2. We can re-use used materials. 3. We should use organic materials such as organic cotton and vegetable dyes. 4. We should implement environmentally friendly production processes.

Fifth step, within our Tools category, we have identified valuable resources that can enhance comprehension of the system and streamline retrospection. To address this matter, our writers have proposed the creation of a standardized eco-label that can be utilized throughout the entire lifecycle of clothing - from the inception of fibres to disposal. It is imperative that these eco-labels are accessible to all and affordable for small and medium-sized businesses.

Presently, there exists a lack of uniformity in the definition and approval process of various ecolabels such as Global Organic Cotton Standard (GOTS), Eco-Tex, ISO-14000, and others. The Second is 12-point leverage system designed by system-thinker Donella Meadows in Slow fashion:

Donella Meadow's 12-point leverage point system is utilized to analyse the fashion industry and identify the "levers and drivers" within the system where a small shift can cause a significant change in the ecosystem. Below are the important findings related to the above points:

3.1 Structure of Information Flows

The information flow in the fashion industry can be categorized into two types: from business to consumer and from business to business.

3.1.1 Business-to-Consumer Information Flow

Previous studies have indicated that consumers may be willing to pay higher prices for "green" products, but it is challenging to communicate the Eco-friendliness of a product to the consumer. Lack of awareness among consumers about environmentally friendly clothing is hindering consumption. Consumers also have a simplistic view of the "greenness" of products. For instance, when competing for the "greenness" of cotton manufacturing versus polyester, the average consumer might assume that cotton is more environmentally responsible because it is a natural fiber, which is not necessarily true. One way for manufacturers and retailers to communicate the Eco-friendliness of products to the consumer is through Eco-labels.

3.1.2 Structure of Information Flow within the Textile Supply Chain

Within the textile and fashion industry, the non-availability of environmentally friendly raw materials is identified as a key constraint in using Eco-friendly raw materials. For example, while India is the world's largest producer of organic cotton, the majority of organic cotton is exported, and it is not easily available locally. Organic raw materials are supplied by a limited number of suppliers and farmers, which restricts the bargaining power of larger retailers.

3.2 Supply Side Constraints on Raw Material Prices and Limited Buffers

The study of leverage points found that the majority of players who are willing to use sustainable materials are constrained by the non-availability of raw materials and high prices. Delayed delivery and high prices of sustainable fiber are also hindering consumption.

CASE STUDY ON DEGRADATION OF WATER QUALITY DUE TO FAST FASHION

Aim of the case study is to see how the fashion industry ranks as the second most polluting industry, contributing 8% of all carbon emissions and 20% of all global wastewater. It is expected that greenhouse gas emissions from the industry will increase by 50% by 2030. To gain a better understanding of the environmental impact of fast fashion, we systematically analysed 65 publications from 1996 to November 2021. There is a growing research interest surrounding fast fashion and water quality. In the last 5 years, 74% of articles were published, with most publications and citations coming from China and European countries. Our summary highlights the evaluation of production processes, including carbon and water footprints, as well as recycling practices to increase the sustainability of the fashion industry. Future research in this growing field should focus on circular economy, social environmental responsibility, and sustainability governance.

Introduction

Excessive waste production poses a serious threat to the environment by compromising air and water quality. The textile and clothing sectors are especially notorious for contributing to this issue by discharging greenhouse gases, generating wastewater, and producing solid waste during production and supply chain processes. This problem stems from social factors such as the emergence of a larger middle class, an increase in women's participation in the workforce, and a surge in individualism, which have given rise to a demand for new clothing styles that reflect new identities. As a result, nearly 90% of the world's clothing production is outsourced to low- to middle-income countries where garments are produced cheaply, with substandard quality, and sold at low prices to facilitate rapid production and consumption. The "fast fashion" model hinges on consumers' desire to keep pace with the latest trends, leading to frequent purchases and premature disposal of clothing. This has earned the fashion industry the dubious distinction of being the second most polluting industry, responsible for 8% of all carbon emissions and 20% of all global wastewater, with a carbon footprint larger than international flights and shipping combined. Furthermore, the fashion industry consumes roughly 93 billion cubic meters of water per year. The confluence of a burgeoning global population and fast fashion has resulted in significant increases in textile production. Per capita fibre consumption rose almost threefold from 1950 to 2008, from 3.7 kg to 10.4 kg per person. Textile fibre production increased by an additional 20.2 million tons to 90.8 million tons from 2007 to 2014, and this number is expected to grow by 3.7% annually. In 2015, 92 million tons of global fashion waste were generated, and this figure is expected to rise by 56 million tons by 2030. While the fast fashion industry has been criticized for creating hazardous working conditions for workers in low- to middle-income countries, it is also vital to address the environmental concerns that arise from this burgeoning industry. The fashion industry consumes vast quantities of water and produces enormous amounts of wastewater, consuming 79 trillion litres of water annually and contributing to approximately 20% of industrial wastewater. Although there is an extensive body of literature on topics related to water treatment options and textile effluents, to our knowledge, a quantitative analysis of the current state of research on the fast fashion industry's role in water quality has not yet been conducted. As a result, our main objective for this study was to systematically review the literature to identify the research already conducted on the textile, garment, and fast fashion industries and their impacts on water quality.

✳ ✳ ✳

According to research, making changes in the consumer behavior can make a huge difference in reducing the environmental harm caused by the fashion industry. It's worth noting that environmental damage doesn't just occur during the production phase of clothing, but also during its use. For instance, washing clothes can release microfibers into the water supply, harming marine life and disrupting their body functions. However, there are ways to reduce your carbon and energy footprint while doing laundry, including using high-efficiency machines, washing at lower temperatures, air drying, using front-loading and full-load machines.

The fast fashion industry has created a culture of disposable clothing, resulting in a significant amount of textile waste that threatens water quality. Despite the decrease in clothing prices, production has doubled worldwide from 2000 to 2015. Social media has become a powerful tool in shaping consumer behavior, with influencers having a significant impact on fashion trends and brands. By using social media to promote recycling and reusing initiatives, we could reduce waste and save water. Additionally, extending the life of garments by just nine months could reduce waste by 22% and save 33% of our water usage.

It is important to recognize the negative effect that fast fashion has on the environment, as its popularity continues to grow for various reasons. Our research shows that there is a growing group of people who are advocating for the fashion and textile industries to address environmental concerns, especially related to water pollution. Although there is already a lot of information available on this topic, more research is needed to fully comprehend the impact of fast fashion on our planet.

The fashion industry has a significant impact on the environment due to its high usage of water and energy, as well as carbon emissions. Although there have been various studies on these issues, there is still a need for further research. Our analysis has highlighted three emerging terms that warrant deeper exploration: "sustainability," "fabric industry," and "water footprint." We suggest performing more investigation into assessing water footprints, promoting recycling initiatives, and developing sustainable practices that support a circular economy for the fashion industry.

Changes in consumer behaviour can also influence the market for fast fashion, leading to more sustainable production processes and stricter regulations on waste disposal. Documenting the negative impact of the fashion industry on the conditions and producing more sustainable practices will be essential to decreasing carbon emissions and wastewater production.

FROM THE AUTHOR

Our mother earth is in danger, and now it is our sole responsibility to save our mother from getting destroyed. Mother Earth has taken care of our basic needs and wants such as shelter to live, clothes to wear, sun for natural heat, and land that provides us to watch beautiful mountains, parks, and animals. Water for cooking food and washing utensils! Air where we can see beautiful birds flying.

Now it is our time to understand, that if we do not fight against climate change, if we do not put a ban on deforestation, if we do not stop illegal poaching and killing of animals for their body parts to earn a profit, if we continue to destroy the environment and overexploit natural resources of our earth, then one day there will be no life that will exist on this beautiful planet.

The fundamental principle of our earth which humans are unaware of is that if they do no start respecting mother earth, then one day the earth has the power to retaliate through Natural disasters by protecting herself by using cyclones, volcanic eruptions, storms, or by emitting diseases like COVID-19 that is an ecocentric virus that took many lives during the period of 2019, and still earth is surrounded with full of diseases that humans might never be able to fight because there will no resources to make the cure. Prepare yourself because now there will be the end of Anthropocentrism and Eco centrism willprevail. Respect mother earth by saving its natural resources and protecting nature, otherwise, this earth will fall into darkness. This book can help you to learn why we should wear sustainable clothes and what are its importance in our society. Tourism is not the best form of travelling from one place to another because it does not help a person to learn about history of the tourist spot, it does not makes us help in learning better moral values like not to pluck flowers, also to keep the natural habitat clean and green. So, we must switch to Eco-Tourism and let us unite together for creating better sustainable future before its too late.

* * *

Mr. Lakshya Mehta is a passout of Amity University, Noida and has started promoting Eco-Friendly Clothes business & being an Evironment saving Enthusiast has
authored this book for environmental information to the masses.

* * *

www.ingramcontent.com/pod-product-compliance
Lightning Source LLC
Chambersburg PA
CBHW062243290526
45794CB00006B/2383